NICOTEXT

AMAZING SPACE

The publisher and authors disclaim any liability that may result from the use of the information contained in this book. All the information in this book comes directly from experts but we do not guarantee that the information contained herein is complete or accurate. The Content of this book is the author's opinion and not necessarily that of Nicotext.

No part of this book may be used or reproduced in any manner whatsoever without the written permission except in the case of reprints of the context of reviews. For information, email info@nicotext.com

Illustrations: Linda Lovén

NICOTEXT part of Cladd media ltd.
www.nicotext.com
info@nicotext.com

Printed in Poland
ISBN 978-91-86283-13-1

My goal is simple. It is a complete understanding of the universe, why it is as it is and why it exists at all.

Stephen Hawking

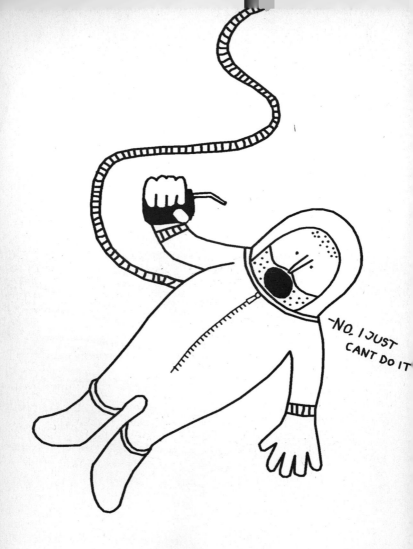

Astronauts cannot burp in space.

41% of the moon is not
visible from Earth at any time.

A 45 kg (100 pound) person on Earth
would weigh 17 kg (38 pounds) on Mars.

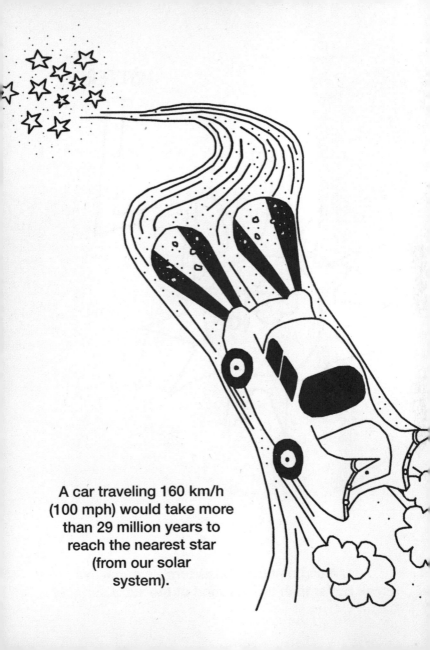

A car traveling 160 km/h (100 mph) would take more than 29 million years to reach the nearest star (from our solar system).

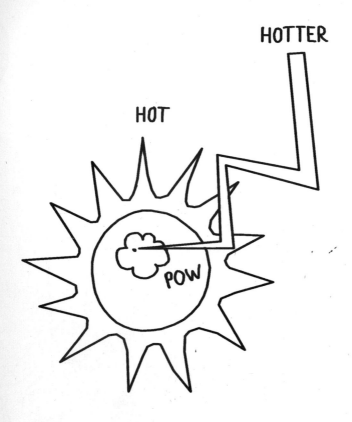

A lightning bolt generates temperatures five times hotter than those found at the sun's surface!

Any space vehicle must move at a rate of
11 km per second (7 miles per second) in
order to escape the earth's gravitational pull.

Approximately 115 tons of ocean salt spray
enters the earth's atmosphere each second.

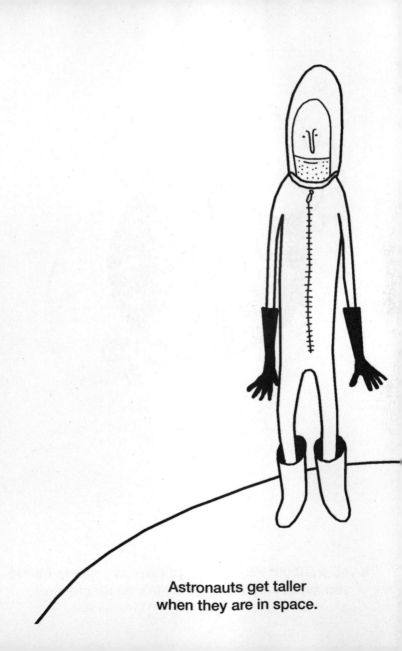

Astronauts get taller
when they are in space.

If you went out into space, you would explode before you suffocated because there's no air pressure.

It takes eight and a half minutes for light to get from the sun to earth.

More than 20 million meteoroids
enter earth's atmosphere every day.

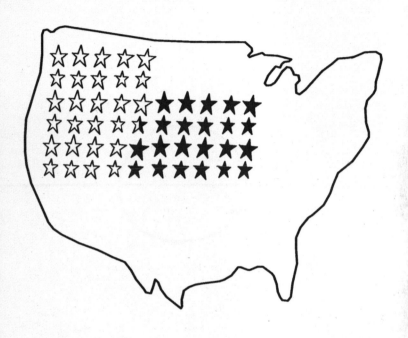

**Only 55% of Americans know
that the sun is a star.**

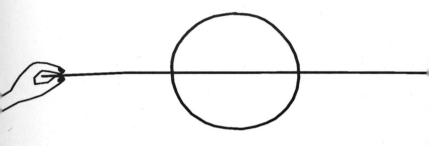

Saturn's rings are about 800 000 km (500,000 miles) in circumference but only about 30 cm (1 foot) thick.

**Some asteroids have other
asteroids orbiting them.**

The earth gets 100 tons heavier every day due to falling space dust.

The earth is hotter
during a full moon.

**The earth travels through space at
1 million km/h (660,000 mph).**

**The earth weighs around
6,600,000,000,000,000,000,000 tons**

The heart of an astronaut gets
  smaller when in outer space.

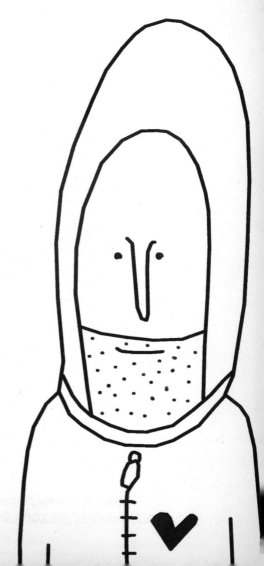

**The moon is moving away from earth
at a rate of 1.5 inches per year.**

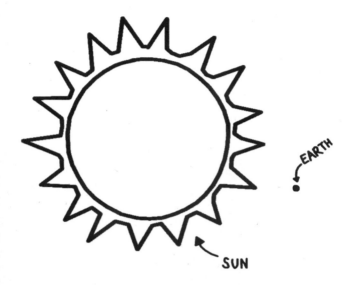

The sun is 330,330 times
larger than the earth!

The temperature of the earth's interior increases by 1 degree Fahrenheit every 18 m (60 feet) down.

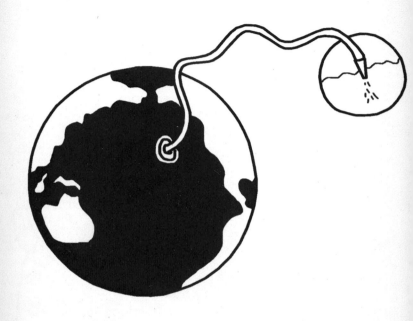

The volume of the earth's moon is the same as the volume of the Pacific Ocean.

There are approximately 3,500 astronomers
in the U.S. – but over 15,000 astrologers.

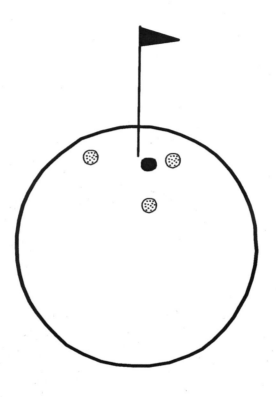

There are three golf balls sitting on the moon.

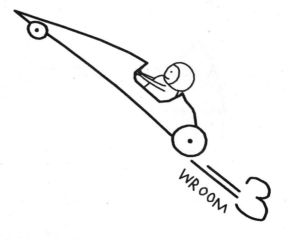

Today's top fuel dragsters take off with more force than the space shuttle.

**Venus is the only planet
that rotates clockwise.**

The average life span of a yellow star, like our sun,
is about 10 billion years. The sun will eventually
burn out in about 5 to 6 billion years.

The Andromeda Galaxy is our closest neighboring galaxy at 2.2 million light years away. It is so bright you can see it with your eyes on a clear evening sky, away from city lights.

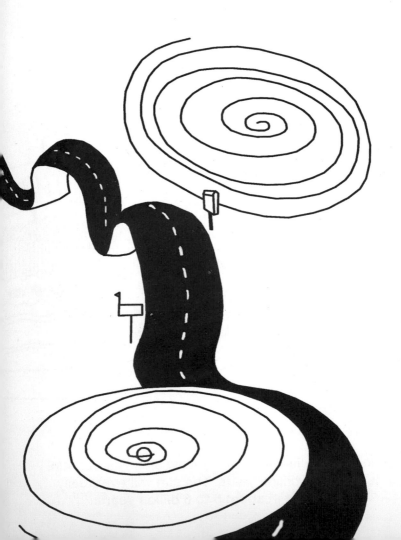

The average temperature on Pluto is
-234 degrees Celcius (-390 degrees Fahrenheit)

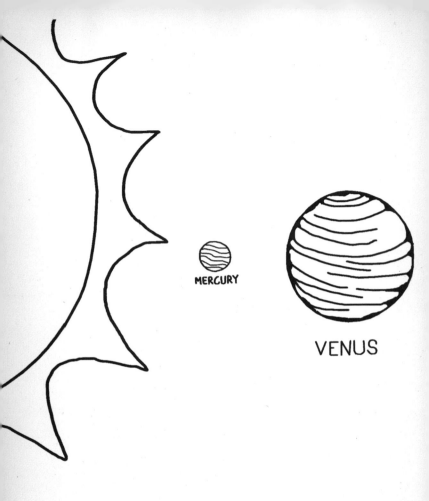

MERCURY

VENUS

The planets Mercury and Venus
are the only planets in our solar
system that don't have moons.

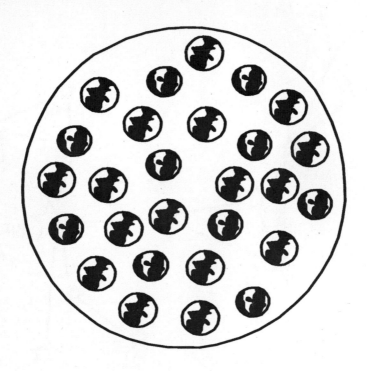

If the planet Jupiter was hollow, you could fit
about 1,400 Earth sized planets inside of it
with a little room to spare.

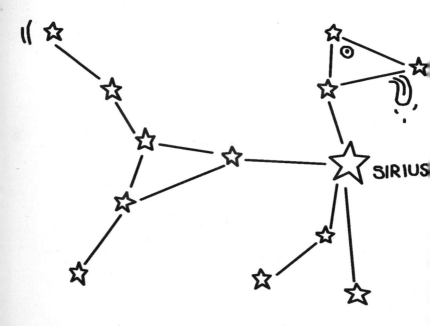

The brightest star in the sky is Sirius.
Also known as the dog star, it is 82 trillion km
(51 trillion miles) from earth or about 8.7 light years.

To reach outer space, you need to travel at least 80 km (50 miles) from the earth's surface.

The planet Mars was named after the Roman god of war. The month of March is also named for him.

For centuries people thought the appearance of a comet was a evil sign that could foretell the coming of plagues, wars and death.

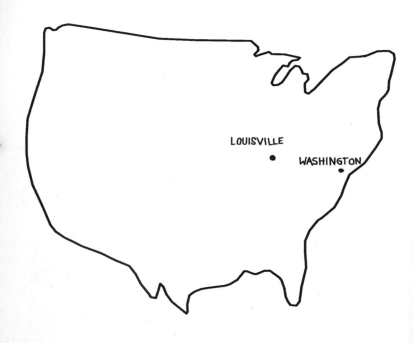

The largest asteroid on record is Ceres. It is so big it would stretch from Washington D.C. to Louisville, Kentucky. A distance of over 1000 km (600 miles)!

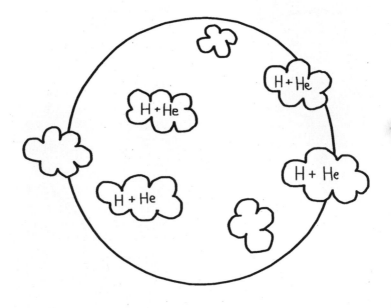

The planet Jupiter has no solid surface,
only layers of gaseous clouds. It is composed
mainly of hydrogen and helium.

METEOR

The average size of a meteor is no
bigger than a grain of sand!

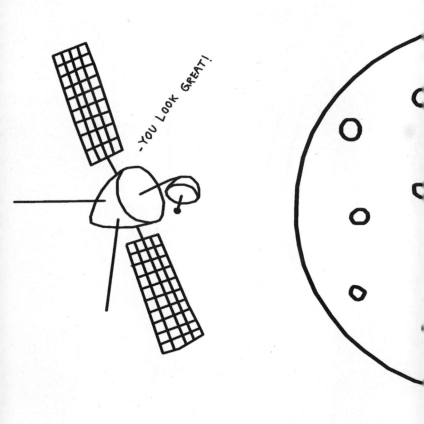

On March 29,1974 Mariner 10 was the first spacecraft to fly by the planet Mercury. It sent back close-up pictures of a world that resembles our moon.

In 1911 a dog was killed by a meteor at Nakhla, Egypt. This unlucky creature is the only one known to have been killed by a meteor.

If the sun suddenly stopped shining it would take eight minutes for people on earth to be aware of it.

HOW TO MAKE A TIMEMACHINE
1. PARK YOUR SPACE VEHICLE
   NEXT TO A BLACK HOLE
2. JUMP INTO THE HOLE AND WAIT
3. TRAVEL TO THE FUTURE

Time slows down near a black hole
and inside ceases to exist.

From space, the brightest man-made place
is Las Vegas, Nevada.

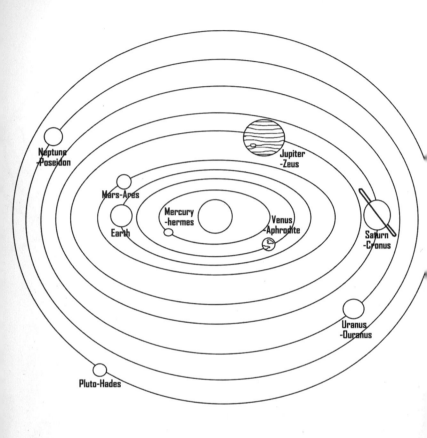

Earth is the densest planet in the solar system and the only one that is not named after a mythical god.

To remember the order of the planets use the phrase 'My Very Eager Mother Just Served Us Nine Pizzas' – the capital letters give the order of planets, beginning with closest to sun – Mercury, Venus, Earth, Mars, Jupiter, Saturn, Uranus, Neptune, Pluto.

In 1958 the US sent 2 mice into space
– Benjy and Laska.

Comets are a mixture of ice and dust
that failed to come together as a planet.
Sometimes they are called "dirty snowballs".

More than 1 million earths could fit inside the sun.
By far, the sun is the largest object in
our solar system.

A Space Shuttle's main engine weighs about
a seventh as much as a train engine but can
deliver as much horsepower as 39 locomotives.

The first "Chimponaut" was 3-year-old Ham,
who rocketed into space on
January 31, 1961.

The footprints left by astronauts on the moon
should last several million years as there is
no wind to blow them away.

Buzz Aldrin's (second man on the moon)
mother's name was Marion Moon.

In space no one can see you cry
– gravity means tears won't flow normally.

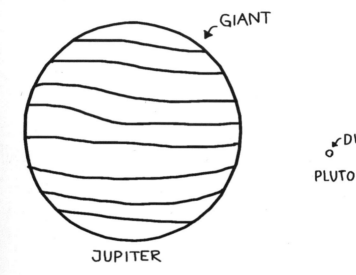

GIANT

DWARF

PLUTO

JUPITER

In 2006, astronomers changed the definition
of a planet. This means that Pluto is now
referred to as a dwarf planet.

The only planet that rotates on its side like a barrel is Uranus. The only planet that spins backwards relative to the others is Venus.

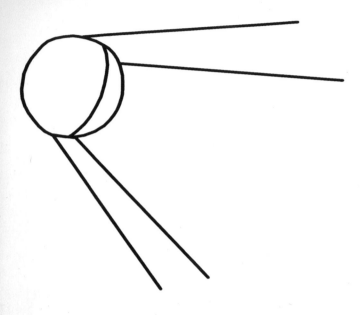

The first man-made object sent into space
was in 1957 when the Russian satellite
named Sputnik was launched.

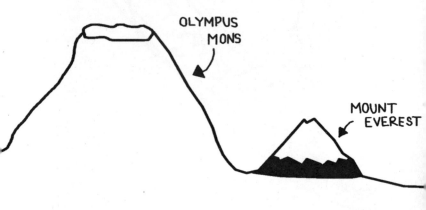

OLYMPUS
MONS

MOUNT
EVEREST

Mars has many massive volcanoes and is home
to Olympus Mons, the largest volcano in our
solar system. It stands 21 km (13 miles) high
and is 600 km (372 miles) across the base.

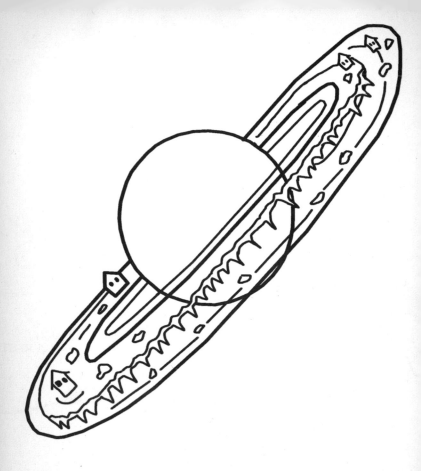

Saturn is surrounded by a system of rings that
stretch out into space for thousands of kilometers.
The rings are made up of millions of ice crystals,
some as big as houses and others as small
as specks of dust.

Saturn is not a peaceful planet. Storm winds race around the atmosphere at 800 km/h (500 mph).

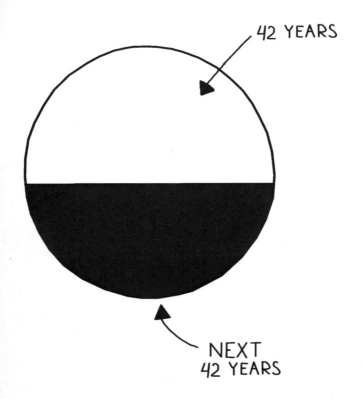

42 YEARS

NEXT
42 YEARS

Since Uranus takes 84 Earth years to go around the sun,
this means that each of its poles is in daylight
for 42 years and in darkness for the next 42.

The ancient Greeks called our galaxy
the Milky Way because they thought it was
made from drops of milk from the breasts
of the Greek goddess Hera.

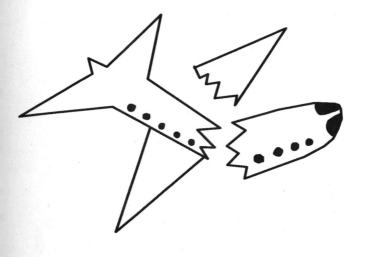

Yuri Gagarin survived the first manned spaceflight but was killed in a plane crash seven years later.

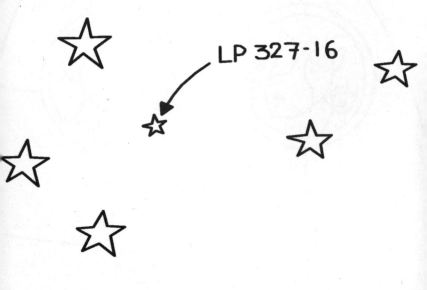

LP 327-16

About 1500 stars are visible at night with
the naked eye in a clear, dark sky. There are
88 constellations altogether. The smallest star
measures about 1700 km (1000 miles) across.
It is a white dwarf called LP 327-16.

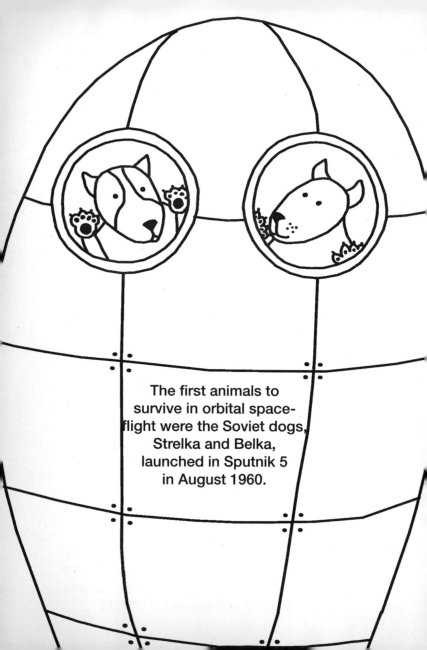

The first animals to survive in orbital space-flight were the Soviet dogs, Strelka and Belka, launched in Sputnik 5 in August 1960.

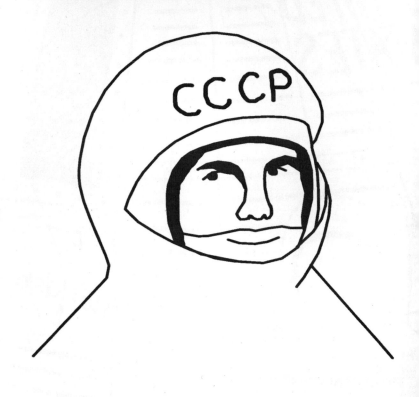

The first person to orbit earth
was Yuri Gagarin, from the USSR,
in April 1961.

UNITED STATES

Friendship 7

The first American to orbit earth was John Glenn in February 1962.

The first woman in space was
Valentina Tereshkova,
from the USSR, in June 1963.

**The first person to walk on the moon
was Neil Armstrong in July 1969.**

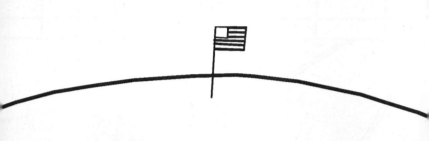

One of the things that Apollo mission did was
deposit a cockroach on the moon. During their
outward flight, the astronauts noticed a cockroach
in their spaceship, but when they returned, the craft
was thoroughly inspected by NASA technicians
and no trace of it was found. The only conclusion
is that it crept out and was left behind.

- ANYBODY HERE?
WHERE DID
EVERYBODY GO?

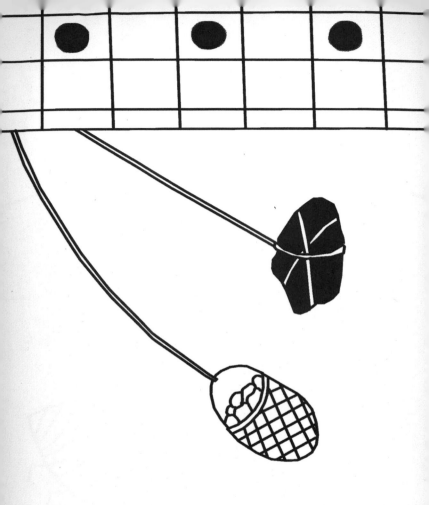

A total of 382 kg (842 pounds) of rock samples were returned to earth by the Apollo and Luna programs.

Astronauts are not allowed to eat beans before
they go into space because passing wind
in a spacesuit damages them.

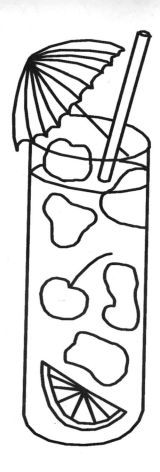

The liquid hydrogen in the Space Shuttle main engine is -423 degrees Fahrenheit (-253 degrees Centigrade), the second coldest liquid on earth, and when burned with liquid oxygen, the temperature in the engine's combustion chamber reaches +6,000 degrees F. (+3,316 degrees C.)

Each of the space shuttle's solid rocket
boosters burns 5 tons of propellant per second.

Comets' tails point away from the sun at all times.
Thus, when a comet is moving away from the sun,
its tail is actually leading. Comet tails are caused by
dust and gas being lost from the comet and then
pushed away from the sun by the solar wind
(charged particles moving out from the sun)
and by radiation pressure from the sun.

An Armageddon meteor came within 450 000 km
(280,000 miles) of the earth on May 21, 1996.
This could have been devastating to life
on earth. Fortunately it missed!

Even before the sun turns into a red giant its increased output will fry the earth, and by the time it turns into a red giant the earth will be a parched ball of bleached rock. But that won't happen for another 5 billion years.

Cold welding is a process when two metals
are stuck together in space. Two pieces of
metal without coating on them will then
begin to form as one.

IN HONOR OF KING GEORGE III OF GREAT BRITAIN

Uranus was also called George's Star.
Sir William Herschel discovered Uranus and
was given the honor to name it George's Star.

The sun loses up to a billion OF grams per second
due to solar winds. Solar winds are charged
particles that are naturally given off by the sun.

**The earth is slowing in rotation 0.002 seconds
per day per century.**

Saturn's density is so low that if you put it
into a glass of water, it would float.

Jupiter's magnetic field is so massive
that it pours billions of Watts into earths
magnetic field every day.

A massive body 100 km (62 miles) wide traveling
at over 512,000 km/h (318 000 mph) crashed into
Mercury to form the Caloris Basin. The impact was
so great it sent shockwaves round Mercury creating
its hilly lineated terrain.

Just a pinhead of the sun's raw material could
kill someone up to 160 kilometers (100 miles) away.

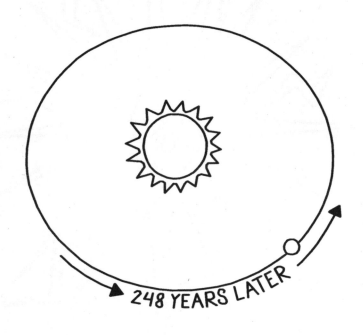

248 YEARS LATER

The length of a Plutonian year is 248 of our years!

A Supernova explosion produces more energy in its first ten seconds than the sun during the whole of its 10 billion year lifetime and for a brief period, it creates more energy than the rest of a galaxy put together.

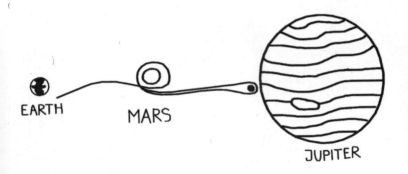

EARTH

MARS

JUPITER

The comet with the longest tail ever recorded is the Great Comet of 1843. Its tail stretched over 800 million kilometers (500 million miles). This is about the same distance the earth is from Jupiter.

The energy in the sunlight we see today
started out in the core of the sun 30,000 years
ago – it spent most of this time passing through
the dense atoms that make the sun.

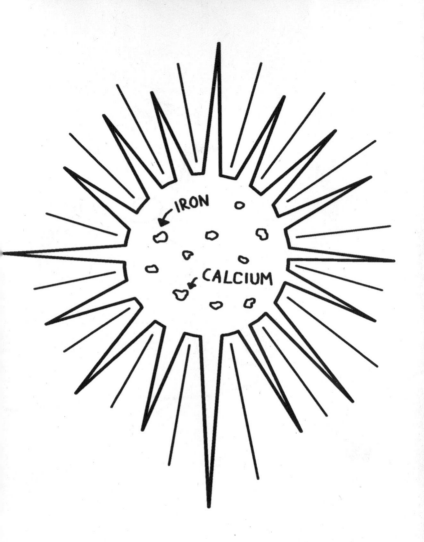

Almost all of the heavier elements in your body
(calcium, iron, carbon) were made somewhere
in supernovae explosions.

Some rocks found on earth
are actually pieces of Mars.

Although the Apollo astronauts did use a special zero gravity pen in the late 1960s, it is an urban myth that NASA spent millions of dollars trying to develop a pen that worked in space while the Russians used a pencil.

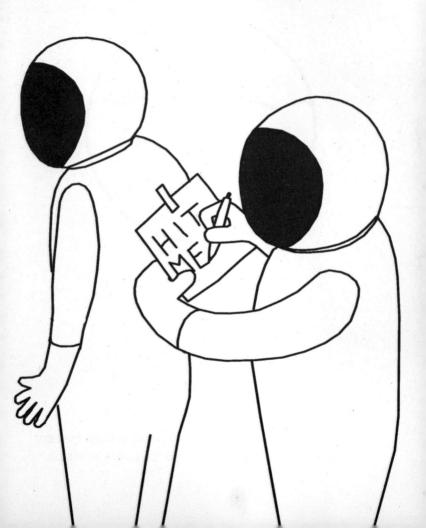

Jupiter's moon Europa may have a liquid water
"ocean" far beneath its ice covered surface.

THE GREAT RED SPOT

The Great Red Spot on Jupiter is a hurricane-like storm system. It is large enough that two earths could fit across it. The Red Spot has been around since at least the early 1600's when it was first detected shortly after the invention of the telescope.

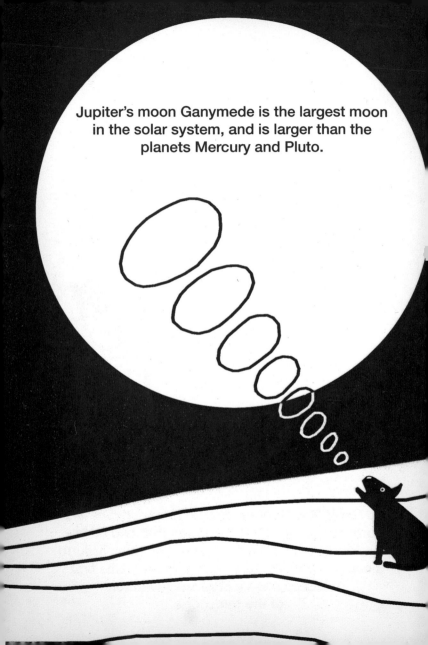

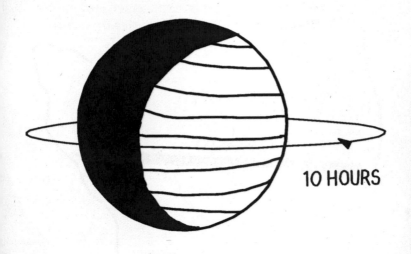

10 HOURS

A Jupiter day, the time required for the planet
to rotate once, is only about 10 hours long.
Jupiter has the shortest day (rotation period)
of any planet in the solar system.

The largest canyon system in the solar system is Valles Marineris on Mars. It is more than 4800 km (3000 miles) long and so would stretch from California to New York. In some places it reaches 4,8 km (3 miles) in depth and 321 km (200 miles) in width.

Many of the larger rocks at the Viking Lander sites on Mars were given names. These included Toad, Badger, and Guppy, all of which were named because of some resemblance to those creatures, as well as ones named for all seven dwarfs, and the largest of the rocks near the landers was named Big Joe.

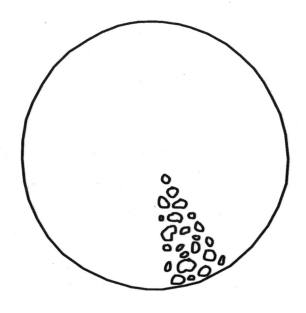

If you added up the mass of all of the thousands of known asteroids in the asteroid belt, the total would be less than ten percent the mass of the earth's moon.

AUGUST 2012

1	17
2 FULL MOON	18
3	19
4	20
5	21
6	22
7	23
8	24
9	25
10	26
11	27
12	28
13	29
14	30
15	31 FULL MOON
16	

A Blue Moon is the second of two full moons that fall in the same month. This can occur because full moon's occur roughly every 29.5 days. A Blue Moon occurs roughly every two and three-quarter years. So, now you know how long once in a Blue Moon really is.

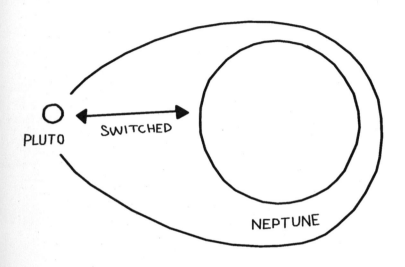

Pluto's elliptical orbit sometimes brings it inside of the orbit of Neptune for a few years. We are currently in one of those periods, so right now Neptune is the farthest planet from the Sun.

Methane gas, which absorbs red light,
is what causes Uranus and Neptune
to look bluish in color.

The atmospheric pressure you would experience on the surface of Venus is approximately equal to the pressure you would experience 3000 feet (approx. 1 km) down in the Earth's oceans, i.e., about 90 times the pressure at the earth's surface.

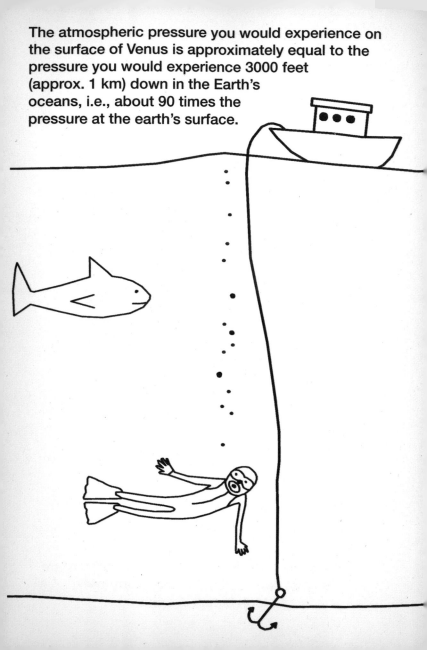

All of the major features on the planet Venus are named after famous women in history and mythology.

Venus is the brightest natural object
in the sky besides the Sun and Moon.
It can be as much as 15 times brighter
than the brightest star (Sirius).

Comet Hale-Bopp is putting out approximately
250 tons of gas and dust per second. This is about
50 times more than most comets produce.

For the first 100 million years or so after the formation of the solar system, a bright, naked eye comet was visible in the skies of earth roughly once a week.

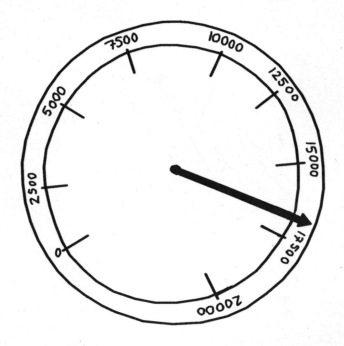

It only takes the space shuttle about 8 minutes
to accelerate to its orbital speed of more
than 27 000 km/h (17,000 mph).

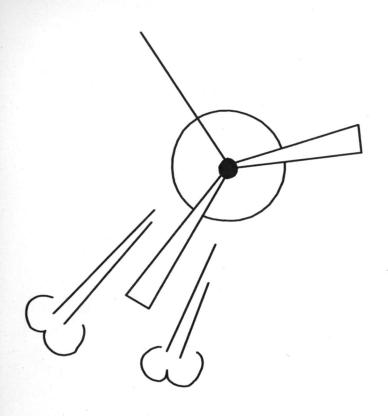

Pioneer 11's speed going past Jupiter was over
172 000 km/h (107,000 mph), the fastest speed
ever traveled by a human-made object.

9 HOURS LATER

At almost 10 billion km (six billion miles) away, Pioneer 10 is the most distant object built by humans. It takes radio signals from earth (traveling at the speed of light: 186,000 miles per second) approximately 9 hours to reach the Pioneer 10 spacecraft, and another 9 hours for the spacecraft's response to reach earth.

A space shuttle and its boosters ready
for launch are the same height as
the Statue of Liberty but weigh
almost three times as much.

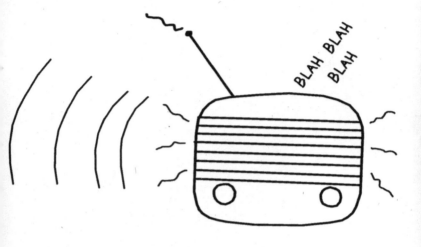

The amount of power transmitted by the Galileo space-craft's radio is about the same amount used by a refrigerator light bulb – about 20 watts. By the time they reach earth, the radio signals from Galileo are incredibly weak (about a billion times fainter than the sound of a transistor radio in New York as heard from Los Angeles).

The Voyager spacecraft delivery accuracy at Neptune was the equivalent of sinking a 3630 km (2260 mi.) golf shot.

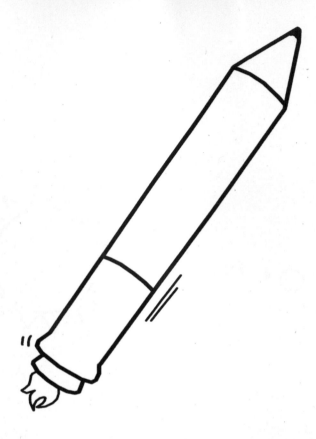

The Solid Rocket Boosters (SRBs) used during space shuttle launches are the largest solid-propellant motors ever flown and the first designed for reuse. Each is 45 m long (149 feet) and 3,6 m (12 feet) in diameter.

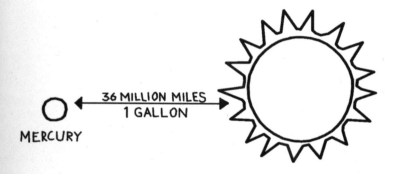

The Galileo spacecraft traveled 3.8 billion km (2.4 billion miles) on its way to Jupiter and along the way used about 253 litres (67 gallons) of fuel to control the flight path and spacecraft attitude. This is the equivalent of about 57 million km per litre (36 million mpg).

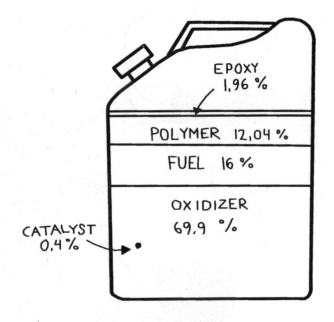

The propellant mixture in each space shuttle Solid
Rocket Booster (SRB) consists of ammonium per-
chlorate, aluminum, iron oxide, a polymer (a
binder that holds the mixture together),
and an epoxy-curing agent.

The only astronaut to have flown into space on
board all five space shuttles is Story Musgrave.

The first American to eat food in space was
Scott Carpenter aboard the Mercury
spacecraft Aurora 7 in 1962.

Only 12 humans have ever visited another world
– all of them walking on the moon during brief
stays between July 20, 1969 and Dec. 13, 1972
as part of the Apollo program.

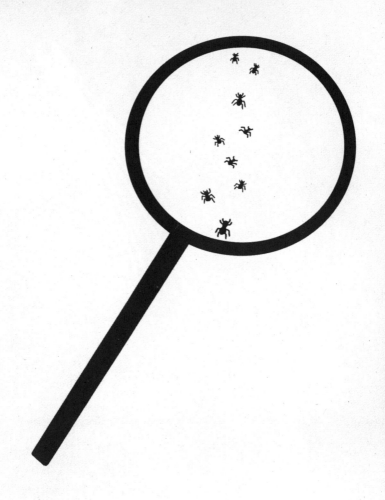

Some of the objects visible in Hubble Space Telescope
images are nearly four billion times fainter than the
limits of human vision.

Comet Hyakutake's orbit will carry it over 1000 astronomical units from the sun before it once again heads back towards the sun in another 7,000 years (1 astronomical unit = the average distance from the earth to the sun = 93 million miles = 150 million km).

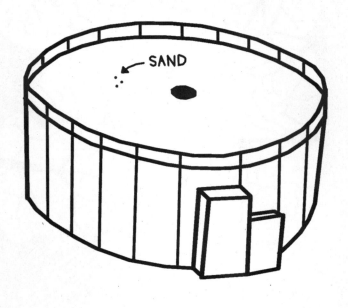

If you suspend three grains of sand in a large sports arena, such as Madison Square Garden in New York, the arena will be more closely packed with sand than our galaxy is with stars.

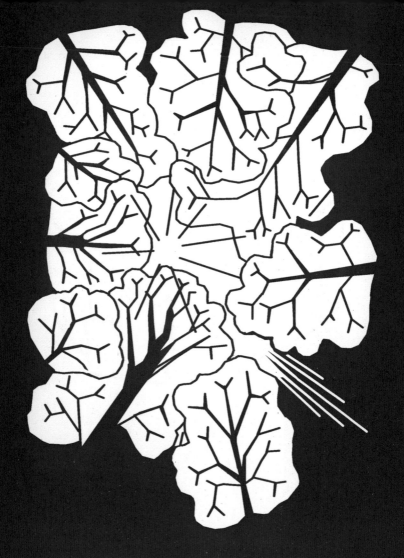

A beam of light travels just over twelve inches in one nano-second (a billionth of a second). Some have suggested naming this unit of distance the phoot.

The elements carbon, hydrogen, oxygen, and nitrogen – all crucial to life – are found in roughly the same proportions in comets and human beings.

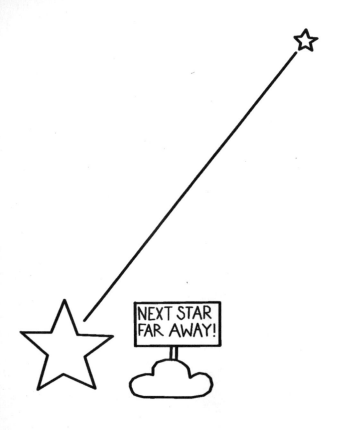

The average distance between stars in the spiral arms of the Milky Way galaxy is currently estimated to be seven light years, or sixty-six trillion kilometers. This distance is equal to roughly 443,000 times the distance between the earth and sun.

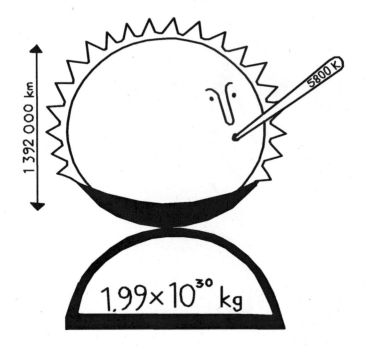

The sun is a fairly average star in terms of mass, temperature, and size.

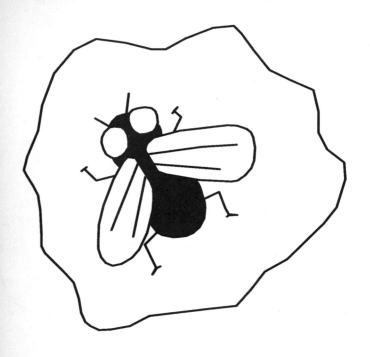

Life is known to exist only on Earth, but in 1986 NASA found what they thought might be fossils of microscopic living things in a rock from Mars.

Oxygen is circulated around the helmet in space suits in order to prevent the visor from misting.

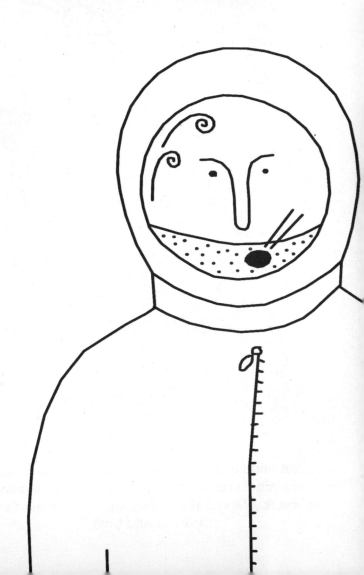

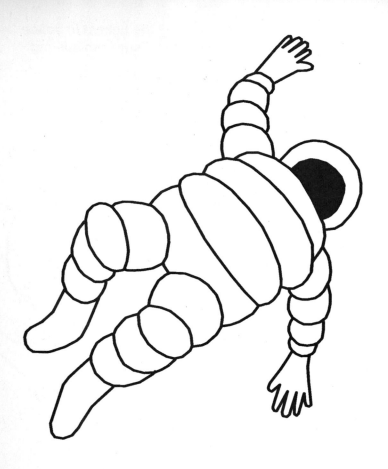

The middle layers of space suits are blown up like a balloon to press against the astronaut's body. Without this pressure, the astronaut's body would boil!

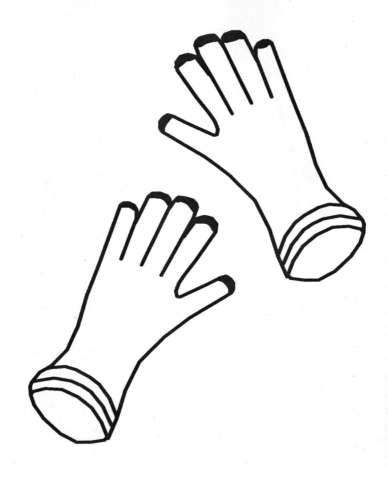

The gloves included in the space suit have silicon rubber fingertips that allow the astronaut some sense of touch.

The full cost of a spacesuit is about $11 million although 70% of this is for the backpack and the control module.

NUMBER OF COMMUNICATING CIVILIZATIONS

EXISTENCE OF LIFE

LENGTH OF SIGNALS SENT IN SPACE

STARS WITH PLANETS

$$N = R^x F_p N_e F_\ell F_i F_c L$$

STARS

LIFESUPPOTING PLANETS

INTELLIGENT LIFE

POSSIBILITY TO SEND SIGNS OF EXISTENCE

The Drake Equation was proposed by astronomer Frank Drake to work out how many civilizations there could be in our galaxy – and the figure is in millions.

SETI is the Search for ExtraTerrestrial Intelligence – the program that analyzes radio signals from space for signs of intelligent life.

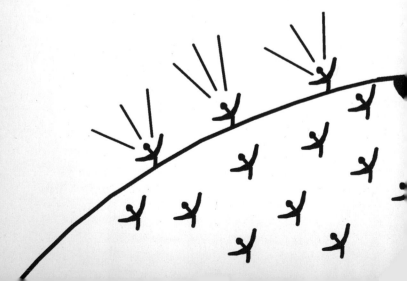

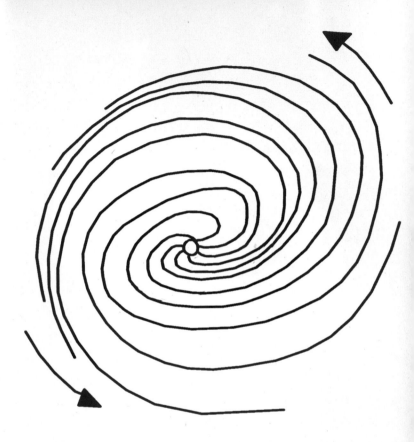

The Milky Way galaxy is whirling rapidly,
spinning our sun and all its other stars at
around 100 million km (62 million miles) per hour.

The sun travels around the galaxy once every
200 million years – a journey of 100,000 light years.

The universe is probably about 15 billion years old, but the estimations vary.

The very furthest galaxies are spreading away from us at more than 90% of the speed of light.

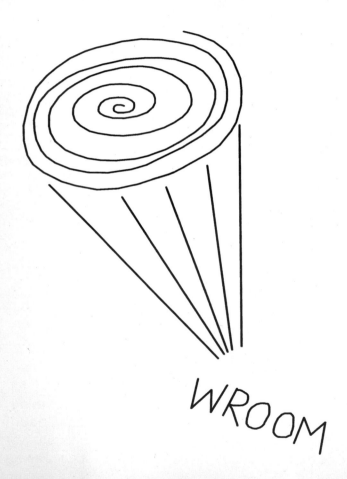

The universe was once thought to be everything
that could ever exist, but recent theories about
inflation (e.g. Big Bang) suggest our universe may
be just one of countless bubbles of space time.

The universe may have neither a centre nor an edge, because according to Einstein's theory of relativity, gravity bends all of space-time around into an endless curve.

**If you fell into a black hole,
you would stretch like spaghetti.**

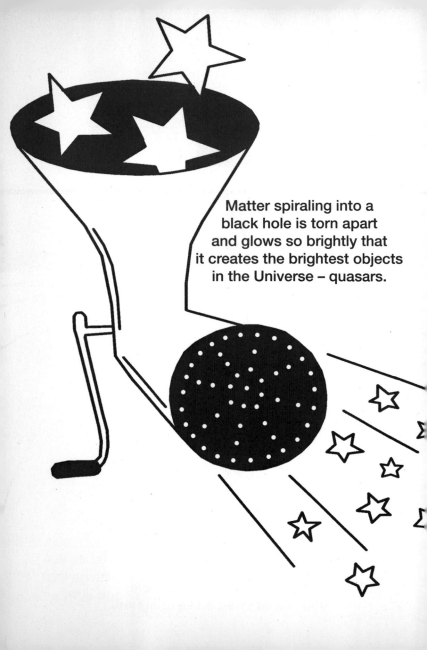

Matter spiraling into a black hole is torn apart and glows so brightly that it creates the brightest objects in the Universe – quasars.

The opposite of black holes are
estimated to be white holes that
spray out matter and light
like fountains.

A day in Mercury lasts approximately as long as 59 days on Earth.

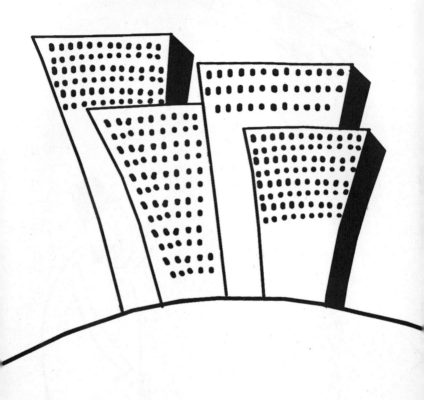

Some particles in Saturn's rings
are larger than skyscrapers.

Halley's Comet appears about every 76 years.

The most dangerous asteroids, those capable of causing major regional or global disasters, usually impact the earth only once every 100,000 years on average.

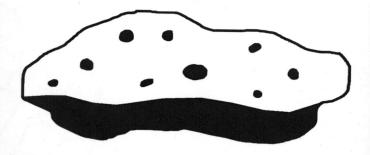

Some large asteroids are so big
they even have their own moon.

THE BOWHEAD WHALE WEIGHTS 60 TONS

The largest found meteorite was found in Hoba, Namibia. It weighed 60 tons.

MILKY
WAY

There are over 100 billion
galaxies in the universe.

**The risk of a falling meteorite striking
a human occurs once every 9,300 years.**

1 MILLION TONS = 166667 ELEPHANTS

A piece of a neutron star the size of
a pinpoint would weigh 1 million tons.

Light reflecting off the moon takes
1.2822 seconds to reach earth.

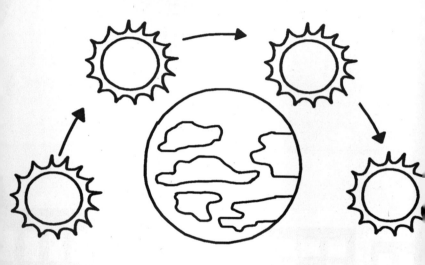

The earth orbits the sun at
107 000 km/h (66,700 mph).

A manned rocket reaches the moon in less time
than it took a stagecoach to travel
the length of England.

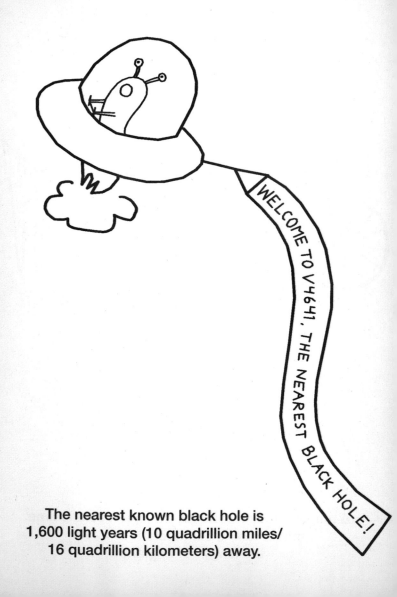

WELCOME TO V4641, THE NEAREST BLACK HOLE!

The nearest known black hole is
1,600 light years (10 quadrillion miles/
16 quadrillion kilometers) away.

4,24 LIGHTYEARS
= 4,01126404×10^{16} METERS

1 METER

The nearest star to our solar system is
Proxima Centauri at a distance of 4.24 light years.

YOU ARE
HERE

Scientists believe that we can only see about
5% of the matter in the Universe. The rest is
made up of invisible matter (called Dark Matter)
and a mysterious form of energy known
as Dark Energy.

The sun produces so much energy,
that every second the core releases the
equivalent of 100 billion nuclear bombs.

OLD STAR

Light from distant stars and galaxies takes so long to reach us, that we are actually seeing objects as they appeared hundreds, thousands or even millions of years ago. So, as we look up at the sky, we are really looking back in time.

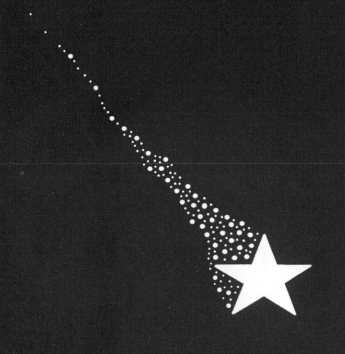

Shooting stars are usually just tiny dust particles
falling through our atmosphere. Comets sometimes
pass through earth's orbit, leaving trails of
dust behind. Then as earth plows through
the dust in its path, the particles heat up,
creating the streaks in the night sky.

The biggest star is VY Canis Majoris,
a red hyper giant 5000 light years
from our solar system. It has been
estimated at around 2000 times the
radius of the sun.

The sun is 4.5 billion years old and produces 383 billion trillion kilowatts of energy.

Sunlight takes 8 minutes to reach the earth and is responsible for the ocean currents and weather patterns on our planet.

The moon is the only non-earth object upon
which a man has walked.

Supernova explosions are capable of
destroying an entire star.

Pulsar, a neutron star that was
discovered in 1967, emits radio waves.

NORTH STAR

Polaris, the North Star, is the only star in the sky that doesn't appear to move from night to night.

There are between
10 sextillion and 1 septillion
stars in the Universe.

**Earth is nearly 149 million km (93 million miles)
away from the sun.**

It takes about 16 million horsepower
to break the earth's gravitational pull.

According to scientists, in around 5 billion years, a day on earth will be 48 hours long.

One light year is roughly 9.46 trillion kilometers
(5,87 trillion miles)!

Our Milky Way galaxy disk is about 100,000 light years in diameter and about 1000 light years thick.

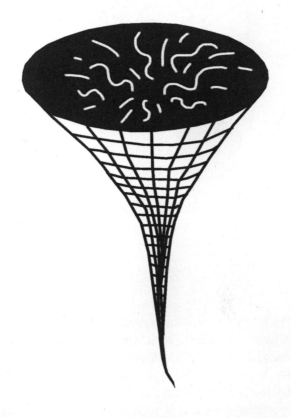

Black Holes are not entirely black. They emit radiation called "Hawking radiation" and eventually disappear.

Pulsars are rotating neutron stars with huge magnetic fields, which emit electromagnetic radiation. They are formed from the core of a star exploding in a supernova. A pulsar is like a lighthouse. When they were first discovered, they were mistaken for a signal from an alien civilization! In fact, the first pulsar source was dubbed "LGM" standing for "Little Green Men".

Quasars are galaxies that travel at a speed
close to the speed of light.

Quasars are the very distant and most energetic galaxies, with an energy output unsurpassed by any other object in the universe. The energy output of a quasar is the same as about 1 trillion suns.

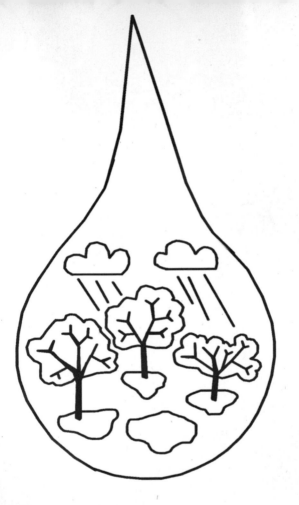

Astronomers have stumbled upon a planet that is quite similar to earth. This new earth-like planet, named GJ 1214b, is bigger than ours, and is more than half covered by water. Nicknamed the "Super Earth", it is around 42 light years away from us, and in another solar system altogether.

An Astronomical Unit is approximately the mean
distance between the Earth and the Sun.
1 AU = 149,597,870.691 kilometers.

An emission nebula is a cloud of glowing gas.

A meteorite is a meteor that has landed on the earth.

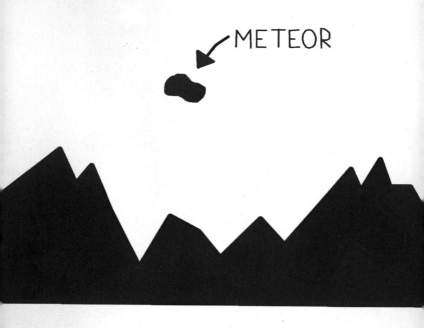

METEOR

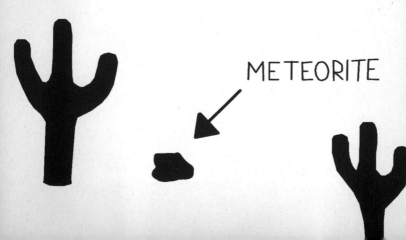

METEORITE

A globular star cluster is a group of
stars that look like a shape of a ball.

The word astronomy is derived from the Greek astronomia. Literally it means "law of the stars".

The name "planet" is derived from the
Greek term planetes, meaning "wanderer".

**Comet Hale Bopp will pass
earth in the year 4937 AD.**

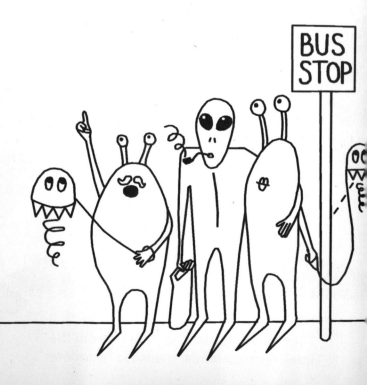

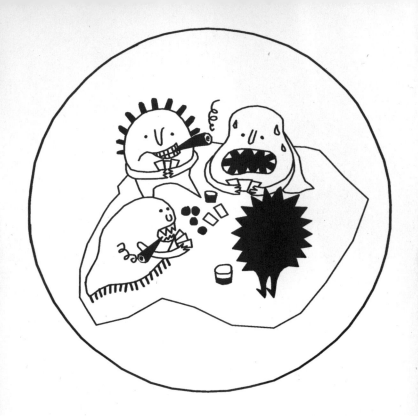

Known as meteor ALH84001, scientists discovered
a meteor in 1996 in Allan Hills, Antarctica that may
possibly be proof of life in outer space. This meteor
fell to Earth 13,000 years ago and tested to be a 4.5
billion year old sample of Mars. Within the tiny cracks
of the meteor, ancient bacteria appear to be present.
Researchers don't know whether this sample actually
represents proof of life in outer space or if it is simply
a contaminated piece of space rock.

The star "Lucy" in constellation Centaurus is
actually a huge cosmic diamond of
10 billion trillion trillion carats.

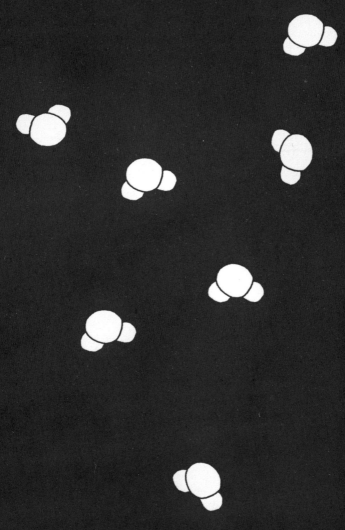

Interstellar space is not complete vacuum:
there are a few hydrogen atoms
per cubic centimeter.

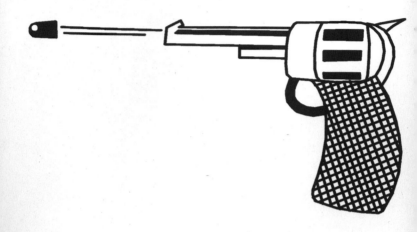

**The Pistol Star is the most luminous
star known – 10 million times the
power of the Sun.**

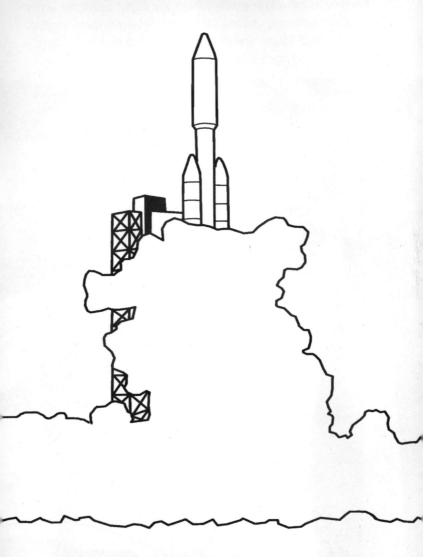

Voyager 1 spacecraft is the farthest
human-made object in the universe.

The sky is blue because when sunlight collides with our atmosphere, colors of the shortest wavelengths (violet and blue) are scattered – and our eyes are more sensitive to see blue.

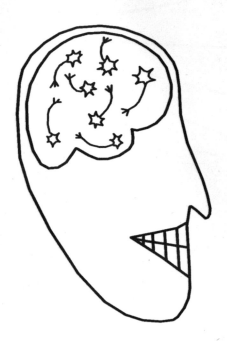

The number of neuron cells in our brain is more than the total number of stars in our galaxy.

The moon is one million times
drier than the Gobi Desert.

A new star is born in our galaxy every 18 days.